Hugo looked around the room. His first day in a new school was about to begin. Many students sat near him, but he had met no one. Hugo felt alone.

Before class began, Hugo saw a student open a letter. Hugo tried to peek down at the stamp. He had never seen one like it before.

Finally, Hugo leaned over to the student. “Hi, my name is Hugo,” he said. “I just came to town seven days ago.”

The boy said, “I am Eli.”

Hugo leaned over even more to point out the stamp. "Are you going to throw that away?" he asked Eli. "Because if you are, I would like to keep it."

"Do you collect stamps?" Eli cried.

"Yes, and how happy you sound!" Hugo remarked.

Eli said, "Come over to my house after dinner, around seven. Bring your stamps."

Hugo came to Eli's house with two stamp albums. He found seven boys and girls there.

"Hugo, meet our stamp club," Eli said.

Hugo met each club member. They sat around a big, round table. Soon they began to study and trade stamps. Hugo never wanted the meeting to end.

Now Hugo no longer felt alone. He felt good because he had pals to collect stamps with. Hugo found out how stamps could help him "lick" his problem!